Filston AMISI KAYUMBA
Adburahim MUNGOMBA SUMAILI

GEOCHIMICAL AND GEOPHYSICAL METHODS FOR THE PROSPECTION OF: Au-Ag-Pt

Filston AMISI KAYUMBA
Adburahim MUNGOMBA SUMAILI

GEOCHIMICAL AND GEOPHYSICAL METHODS FOR THE PROSPECTION OF: Au-Ag-Pt

ScienciaScripts

Imprint
Any brand names and product names mentioned in this book are subject to trademark, brand or patent protection and are trademarks or registered trademarks of their respective holders. The use of brand names, product names, common names, trade names, product descriptions etc. even without a particular marking in this work is in no way to be construed to mean that such names may be regarded as unrestricted in respect of trademark and brand protection legislation and could thus be used by anyone.

Cover image: www.ingimage.com

This book is a translation from the original published under ISBN 978-620-6-71672-3.

Publisher:
Sciencia Scripts
is a trademark of
Dodo Books Indian Ocean Ltd. and OmniScriptum S.R.L publishing group

120 High Road, East Finchley, London, N2 9ED, United Kingdom
Str. Armeneasca 28/1, office 1, Chisinau MD-2012, Republic of Moldova, Europe
Printed at: see last page
ISBN: 978-620-7-85885-9

GEOCHIMICAL AND GEOPHYSICAL METHODS FOR THE PROSPECTION OF Au-Ag-Pt and U

Geologist Filston AMISI KAYUMBA

I. INTRODUCTION

Prospecting is a discipline of geology that deals with the search for new mineral deposits (anomalies).

It therefore requires the use of a number of methods of which those who carry it out are sometimes unaware, and which lead to mediocre results after the efforts made.

Nevertheless, in the field of earth science (geology), there are now simple methods proposed by scientists that can be made available to communities in order to overcome these obstacles.

This is the case, for example, with geochemical and geophysical methods that may be developed later as part of our research.

Thus, in this study we will be called upon to answer the fundamental question of which geochemical and geophysical methods can be used for prospecting: Au-Ag-Pt-U in order to provide indications on the internal mineralogical composition of the earth.

Faced with this situation, we asked ourselves the following questions:

- ❖ What geochemical and geophysical methods should be used to prospect for the aforementioned mineral substances?

- ❖ how can these methods be used to describe and highlight underground anomalies?
- ❖ How can this knowledge be passed on to the target population?

To help us in our quest, we have set ourselves the following objectives:

- ➢ Identify Geochemical and Geophysical methods to explore Au-Ag-Pt-U.
- ➢ Demonstrate the principles, applications and techniques of each method in order to discover underground anomalies.
- ➢ To popularise methods of investigating substances among communities, scientists and mining researchers specialising in the earth sciences for prospecting: Au-Ag-Pt-U, while providing the theoretical bases of geochemistry and geophysics as well as the principles, techniques and applications relating to each method used.

As a provisional response to these concerns, we believe that :

- ➢ Lithological methods, soil geochemical methods, stream geochemical surveys and heavy mineral methods are the geochemical methods to be used in prospecting for these ores, while seismic, gravimetric, electrical and magnetometric methods are referred to as geophysical methods;
- ➢ The description and implementation of subsurface anomalies are made possible by demonstrating the principles, applications and techniques of each of the above methods;
- ➢ This knowledge will be passed on to the target population through sessions that explain the theoretical bases of geochemistry and geophysics, as well as the principles, applications and techniques of each method.

II. METHODOLOGICAL

To achieve this, we will use both analytical and descriptive methods. The former has been useful in analysing the situation in order to understand what the communities are doing in the field of prospecting, which is the crucial stage in the exploitation of mineral substances, while the latter will help us to describe the methods to be used to carry out purely scientific prospecting.

These two methods were supported by documentary analysis and interviews as techniques for gathering the data required for this research project.

Through documentary analysis, we consulted a number of works related to our research theme, and through interviews, we were able to share with members of certain communities interested in mining.

I. RESULTS

At the end of the surveys carried out in the field, and with reference to the objectives of our study, we have come to the conclusion that the following principles and applications are very useful for carrying out good prospecting, thus making it possible to describe and apply subsurface anomalies:

a. **Geochemical methods**

- **LITHOLOGICAL METHODS**

Principle

Lithological methods are based on the identification of primary dispersal aureoles of deposits or ore bodies as well as their secondary aureoles and dispersal trains in : Au-Ag-U-Pt.

Application and technique

A litho-geochemical survey is generally combined with a cartographic survey. The structure and rock types of the outcrop visited are systematically described. Homogeneous and representative geological litho samples are taken from the outcrop visited. Samples are generally taken from the outcrop except where the outcrop contains several rock types, in which case all the rocks in the samples are sampled. The sample taken is generally the size of a fist and we ensure that it is free of any surface alteration.

The bedrock on which litho geochemical studies are carried out may be altered as a result of numerous physical and/or chemical processes. Litho geochemistry will therefore no longer be the method of choice for surveying the minerals resulting from this alteration. Instead, ground-based geochemical survey methods are used.

➢ GEOCHEMICAL SOIL SURVEY METHODS

Principle

Soil is the thin layer of topsoil that continuously covers the surface of the earth. Soil is the result of weathering processes linked to climate, bedrock and life. As it evolves, a soil thickens and differentiates into successive distinct and often irregular layers known as horizons. When the redistribution of elements is significant, a soil profile is differentiated into an eluvial horizon (depleted) and an illuvial horizon (enriched) in Au-Ag-Pt.

In exploration geochemistry, soil geochemical survey methods for : Au-Ag-Pt and U. The different soil horizons are sampled in an attempt to detect the presence of Au-Ag-Pt and U, as the behaviour of metals depends on this. Horizons are chosen according to the metal of interest, i.e. : Au-Ag-U-Pt metals in different horizons are shown below:

- Horizon H: the humus in this horizon may concentrate certain elements such as copper, nickel, lead, zinc, etc.
- Horizon A: even if it is leached, it is sometimes very rich in gold (Au) because it is geochemically not very mobile.
- Horizon B: in which mobile metals such as lead and arsenic are released after only a slight chemical attack.
- Horizon C: anomalies in this horizon, the parent material of the soil, are often detrital in origin.

Application and technique

Geochemical prospecting for : Au-Ag-Pt-U soil geochemical survey methods, small areas of 1km² to 5km² are prospected. A potential area is first circumscribed by other geochemical techniques, such as stream geochemistry, or by geophysical methods. Geophysical surveys often reveal numerous anomalies.

Ideally, it would be too costly to drill holes over each anomaly. Geochemistry is perfect in these circumstances; it is cheaper to take soil samples over each geophysical anomaly than to drill over each one.

Where there is no geophysical signature, soil geochemistry is still used in the following two cases:

1. Gold-bearing valves whose sulphide content is too low to be detected by geophysical methods.
2. Stratiform deposits are geophysical signatures.

Finally, we describe the site - vegetation, topography, human settlements, boulders, etc. - and note anything that might have an influence, however slight, on the content of the sample.

Once the details of small areas have been worked out, the geochemical soil survey will give way to another geochemical method called **CREEK GEOCHIMICS.**

- **METHOD OF STREAM GEOCHEMISTRY**

Stream geochemistry involves the search for Au-Ag-Pt-U in stream water and the sediments it carries.

Principle

The sediments of the streams of materials arranged according to their degrees of mobility as described in the table of stability relative to the alteration of certain minerals in a source: Andrews-Jones, 1968 and Perel Man, 1977, in Peter, 1987.

The elements likely to give rise to anomalies in stream sediments take the form of :

1) Primary minerals such as gold, cassiterite, etc., which resist alteration without forming, are dense and travel with the coarse fraction of sediments.
2) Alteration minerals are for the most part reliable minerals, and are quickly found as grains in sediments. They then tend to remain suspended in water. Only some of them are incorporated into the clay fraction of the sediment.
3) Precipitates from river water. These can be trace elements or others, these minerals coat the grains of the detritic fraction of the sediments or tend to remain in the water in suspension and not in dissolution. The contribution of this form of element is generally small.
4) Absorption by clays. A balance is established between the sediment and the water. At the sampling site, the concentration of a given element in the sediment, i.e. : Au-Ag-Pt-U reflects that of the water.
5) Organic material that has incorporated sediments. This form includes plant debris growing near the mineralised A horizon and the organic matter responsible for anomalies, organic matter formed by biological activity in the watercourse.

Application and technique

Sectors with young and active drainage are generally chosen where watercourses transport large quantities of sediment, since in old and not very erosive sectors, the metalliferous content tends to homogenise and therefore form a background noise that misses anomalies. It is therefore essential to establish the type of coverage and, above all, the sampling density required to reach the Au-Ag-Pt-U target. The density depends on the drainage network.

A sample is taken from the active part of the stream underwater, and as only the fine, siliceous fraction is analysed, the sample must contain as many fine particles as possible.

In many cases, particularly Au-Ag-Pt-U when the current is too fast, the quantities of fine fractions are low, so it is a good idea to sample around blocks or in holes where the water is calmer. A quantity of material of around 100g is sufficient for analysis, and the sample taken is placed in a numbered kraft paper bag. Small quantities are generally taken from several locations a few metres apart.

Finally, it is important to write down the location, noting :

- Water depth ;
- The width of the watercourse ;
- Current speed ;
- Sediment grain size ;
- The percentage of organic matter.

Stream sediment geochemistry is not the only method for discovering Au-Ag-Pt-U concentrations in streams.

➢ GEOCHEMICAL METHODS FOR SURVEYING HEAVY MINERALS

Principle

The aim of heavy mineral geochemistry is to discover high density minerals. It applies to most common metallic minerals, which are generally of above-average density.

The following table shows some minerals according to their respective densities: relative density of minerals.

MINERAL	DENSITY
Visit	19,3
Ag	10,5

Source: Maurice and Mercier, 1985 cited in La prospection minière, pp160.

Application and technique

Sampling for heavy minerals (Au-Ag) is carried out manually using a pan or mechanically using a dredge. The sampling site is chosen on the basis of two main criteria:

The distance from possible sources of contamination and the possibility of recovering a good quantity of Au-Ag for a given volume of sediment.

In a fairly fast-flowing stream, once the sampling site has been chosen, proceed as follows:

You take a drum and fill it to about three-quarters of its capacity, removing coarse material either at home or using a mesh screen, as the finer the material, the better it concentrates. We then set up in a place with a moderately strong current sufficient to carry away the reject material.

You push the pan underwater and break up the alluvium by passing your hand through the pan. The contents of the pan are then put into suspension using a vigorous gyratory movement, still underwater. The Au-Ag tends to sink to the bottom of the pan and the light minerals remain in suspension. The pan is tilted to evacuate the light minerals, but the pan must not be tilted too far for fear of the heavy minerals escaping. This process continues until only a few grams of the dark minerals remain in the pan. The products of the previous operations are placed in a plastic bag and the process is repeated until approximately 40g of concentrated material is obtained. It generally takes 4 to 5 concentrates to achieve this result. Manual sampling has the advantage of being economical and requiring very little equipment, but it is not suitable for treating large quantities of alluvium. For this purpose, slice dredging can also be used.

The Sluice dredge enables a large volume of sediment to be processed relatively quickly by deep suction, which also increases the chance of collecting heavy minerals (Au-Ag). The result is a sample with a high concentration of heavy minerals.

In addition to all the geochemical methods mentioned above, prospecting for Au-Ag-Pt-U also requires other methods in order to pinpoint the mineral substance shown.

b. Geophysical methods

- **SEISMIC METHODS**

Seismic prospecting is a prospecting method used in : Au-Ag-Pt-U, which makes it possible to visualise geological structures at depth by analysing seismic wave echoes.

The seismic waves studied can have natural causes (earthquake or quake) or artificial (vibrating truck, explosive, air truck, etc.) which is why we have the following seismic waves:

a) Volume waves

They are applied in the theory of elasticity and solid mechanics. They propagate inside the globe according to laws similar to those of geometrical optics. Volume waves are subdivided into :

- Primary waves (p): also known as compressional waves or longitudinal waves, propagate in all media. The displacement of the ground that accompanies their passage occurs by successive expansion and compression, parallel to the direction of wave propagation.
- Secondary waves: also known as shear waves or traversable waves. When they pass through, the ground moves perpendicular

to the direction of wave propagation. These waves do not propagate in liquid media.

b) "S" surface waves

They are guides on the earth's surface. Their effects are similar to the ripples formed on the surface of lakes. They are slower than volume waves, but their capacity is generally stronger and they concentrate the maximum amount of energy.

A distinction can be made between :

- Love wave: propagates only in non-homogeneous solids. The through wave is polarised in the horizontal plane and results from constructive interference between horizontal waves.
- The Rayleigh wave: propagates near the surface of homogeneous and inhomogeneous media. The displacement is complex, quite similar both horizontally and vertically.
- The p-wave: which alternately compresses and stretches the rocks. The vertical component of the seismograph is recorded.
- The S wave: propagates by shearing the rocks laterally at right angles to its directions of propagation. It is well recorded on the horizontal components of seismographs.

In all cases, they follow the same laws of propagation as light waves. The three main seismic prospecting techniques used in this work are :

1) acoustic logging

It uses the direct transmission of waves to measure the speed of sound in the rock traversed by drilling.

2) **reflection seismic**

*** principle of reflection seismic**

Reflection seismic studies the reflection of seismic waves at the interfaces between several geological layers. It provides a 2D or 3D image of surfaces typically of the order of 1000 km² at depths of less than 10 km.

The seismic survey is divided into 3 main stages:

- seismic data acquisition ;
- data processing ;
- interpretation.

Compressed air cannons are used at sea, vibrator trucks or dynamite on land to create a wave that will propagate underground. The wave created is either an impulse (airgun dynamite), or a sinusoid whose frequency varies over a defined spectrum over a period of time (vibrator truck used wherever the terrain allows.

A number of parameters characterise a seismic survey in Au-Ag-Pt-U prospecting, the main ones being :

- The size of the 2D or 3D seismic profile ;
- Spacing between sources and sensors (bin size) ;
- Maximum coverage (number of times an area of the sub-clear)

The seismic data then undergoes complex computer processing.

3) **seismic** *refraction*

Refraction seismic uses wave propagation along the interfaces between geological levels. This method is particularly suitable for certain civil engineering and hydrology applications. It can be used to estimate the velocity and dip of layers. In practice, it is currently limited to targets at depths of less than 300m.

➢ ELECTRICAL METHOD

The electrical method is based on the surface measurement of the intensity and the potential difference between the different electrodes of the measurement dissipative. The ratio of these two parameters is used to calculate the resistivity of the underlying ground. There are a large number of such devices, which fall into the following two groups:

- Electric boreholes: for vertical exploration and,
- Electric trains: for horizontal exploration.

These methods have current sources and wiring, electrodes, thousand volts,... are the devices they use. Having talked about the methods for measuring induced physical phenomena, let's move on quickly to talk about natural physical phenomena.

➢ GRAVIMETRIC METHOD

The purpose of this method is to determine, from the surface of the ground, the differences in density of the subsoil by measuring local variations in the field of gravity.

Principle

If we imagine the surface of the Earth as a perfect, homogeneous sphere. The acceleration of gravity "g" at each point on this surface would be constant and equal to 9.81 m/s^2. But since the earth is neither spherical nor homogeneous, the actual values of 'g' measured at the surface of the earth are variable and never correspond, except accidentally, to this theoretical value, since many factors modify it. These include latitude, topography and terrain density. As the variations in "g" caused by differences in soil density provide information about the make-up of the subsoil, it is important to isolate them and, to do this, to correct the measurements of "g" taken on the ground for the variations produced by each of the other factors (correction for altitude, correction for topography, etc.).

Measuring equipment: there are four types of measuring equipment used in gravimetry: torsion balances, pendulums, gravimeters and vibrating ropes.

➢ MAGNETOMETRIC METHOD

Magnetometric prospecting uses variations in the Earth's magnetic field caused by differences in the magnetic susceptibility (the ability of a body to hold itself together) of rocks to identify underground formations. Magnetometry is of interest for Au-Ag-Pt-U prospecting because the presence of rocks with high magnetic susceptibility in the subsurface distorts the lines of force of the magnetic field, causing anomalies that can be drawn as isonometric curves on maps. This distortion depends not only on the susceptibility of the rocks themselves but also on the depth at which they are located.

The instrument used in magnetometry is the magnetometer, which was originally mechanical but since 1940 has been replaced by the electric magnetometer.

GENERAL CONCLUSION

Throughout this work, we have talked specifically about mineral prospecting methods. We began by illustrating the geochemical prospecting methods of : Au-Ag-Pt-U, where we explained what geochemical prospecting is, its principles, applications and techniques. Geochemical prospecting enables geochemical anomalies to be identified by systematically measuring the content of one or more trace elements in layers, soils or in the sediments of streams, lakes and rivers. Geochemistry uses several methods. However, we have only presented four of the most common: the lithological method, which consists of a systematic description of the structure and types of rocks containing trace elements. The soil geochemical survey method, which involves sampling and geochemical analysis of different soil horizons to detect signs of mineralisation. We have also described the geochemical survey methods for streams and heavy minerals. The latter two methods involve taking samples of stream sediments, as river water carries minerals and deposits them according to their degree of mobility. These two methods are similar, the only difference being that in heavy mineral geochemistry we look for minerals with high densities.

Secondly, we discussed geophysical prospecting methods, including gravimetry and magnetometry for natural physical phenomena. The purpose of gravimetry is to determine, from the surface of the ground, the different densities of the subsoil layers in order to measure local variations in the field of gravity, while magnetometry is used to identify subsoil formations and variations in the Earth's magnetic field caused by differences in the magnetic susceptibility of rocks. For induced phenomena, we have described seismic and electrical methods. Seismic methods consist of inducing seismic activity at a point in the ground and measuring the time taken for the waves thus created to travel through varying thicknesses of the ground and reach the ground surface as a result of reflection or refraction. Electrical methods are based on surface measurements of intensity and potential difference.

We therefore feel that we have achieved our objectives, which were to provide geological researchers and scientists in general with a basic understanding of geochemical and geophysical prospecting methods and to provide answers to a number of related problems. We recognise, however, that we have not solved all the problems of geochemical and geophysical prospecting, but we suggest that all those with a scientific curiosity should take up the work and extend the studies in one aspect or another.

BIBLIOGRAPHY

A .WORKS :

- ✓ CHAISSIEER JB, MOREER ; Manouel du prospecteur minier,Edition du BRGM ?Paris ,1980.
- ✓ DIPLESSIS D, et all, prospections minieres, Module Editeur, Québec 1960.
- ✓ MELCHIOR A. et All "Mineral exploration using soil geochemistry in the AKTEAGA region, Michoacán, Mexico".
- ✓ DICTIONNAIRE DE GEOLOGIE, Paris, Masson, 1984.
- ✓ ENCYCLOPEDIA UNIVERSALIS ? France SA, Encyclopedie Universalis.

B. COURSES STUDIED :

- Professor NZOLANGI. (2018) General Geochemistry notes.

- CT SYVIHOLIA C. (2018) notes in **General Geophysics.**

D .WEBOGRAPHY

1. www.aiea.com
2. www.Google. En
3. www. Onlym.coms

CONTRIBUTION TO THE PETROGRAPHIC AND STRUCTURAL STUDY OF NTAMBUKA/IDJWI (SOUTH KIVU, D.R.C.)

Filston AMISI KAYUMBA - MUNGOMBA SUMAILI Adburahim

SUMMARY

Idjwi is one of the largest islands in the African lakes. This stretch of land lies in Lake Kivu, in the western branch of the East African Rift system. The petrographic and structural study of this western part of the East African Rift has enabled certain geological formations to be determined and certain brittle deformations (diaclase, joint) to be oriented on the basis of tectonic parameters such as the orthogonal principal stress tensor ($\sigma1$, $\sigma2$ and $\sigma3$), frequency rosettes and plane pole dispersion diagrams. These parameters were obtained using Win-tensor and dips software from plane orientation data of filled fractures and miscellaneous fractures measured during a field survey campaign.

On the basis of this approach, the geological formations of this zone are respectively of the Ruzizi sub-group such as: kaolins, silicified limestones, pyroclastites, schists, micaschists, granites, pegmatites and dioritic rocks.

Continuous deformation is rare, and the few planes of schistosity run NE-SW. The diaclases give a single structural network ranging from 140° to 160° with a preferential direction of N335°E/48°ENE, which are derived from compression with $\sigma1$: N106°E/0°WNW, $\sigma2$: N15°E/79°SSW and $\sigma3$: N196°E/11°NNE. For the veins, the preferred direction is N190°E/55°WNW. Originating from a distension of which $\sigma1$: N166°E/68°NNW, $\sigma2$: N352°E/22°SSE and $\sigma3$: N261°E/2°ENE.

Key words: Kivu, Idjwi, Rift, tectonics, deformation and Ruzizian.

ABSTRACT

Idjwi is one of the largest isles in African lakes. This tract of land is found in Kivu Lake, in the western branch of the East African Rift system. The petrographic and tectonic study of this western part of the East African rift made it possible to determine various geological formations and to orient some brittle deformations (diaclasis, joints) based on the tectonic parameters such as the tensor of the orthogonal main stresses ($\sigma1$, $\sigma2$ and $\sigma3$), the rosettes of the frequencies as well as the dispersion diagrams of the poles of planes.

These parameters were obtained using Win-Tensor software and dips from the orientation data of the planes of the completed breaks and the various fractures measured during a survey campaign.

On the basis of this approach, the geological formations of this zone are, respectively, of the Ruzizi sub-group such as: kaolins, silicified limestones, pyroclastites, shales, micaschists, granites, pegmatites, dioritic rocks.

Continuous deformations are rare; bedding planes are oriented around NE-SW. The diaclases gives a single structural network which goes from 140 ° to 160 ° with main direction of N335 ° E / 48 ° ENE which are resulting from a compression with $\sigma1$: N106 ° E / 0 ° WNW, $\sigma2$: N15 ° E / 79 ° SSW and $\sigma3$: N196 ° E / 11 ° NNE. For the veins, the preferred direction is N190 ° E / 55 ° WNW. From a distension of which $\sigma1$: N166 ° E / 68 ° NNW, $\sigma2$: N352 ° E / 22 ° SSE and $\sigma3$: N261 ° E / 2 ° ENE.

Key words: Kivu, Idjwi, Rift, tectonics, deformation and Ruzizian.

I. INTODUCTION

The DRC's subsoil is one of the richest in the world in terms of geology and mineralisation.

The DRC has deposits containing around fifty minerals, but only a dozen of these are mined: copper, cadmium, diamonds, gold, tin, tungsten, manganese and a few rare metals such as Coltan (Ta). (www.ambardcprague.com).

In order to do this, studies carried out in the past must always be updated by various research projects, including dissertations, post-graduate work, various scientific publications, mining exploration work, etc.

It is with this in mind that in this work we make a moderate contribution to the geological knowledge of our country, more particularly the petrographic and structural study of the formations of the NTAMBUKA chiefdom in the southern part of the Idjwi territory in South Kivu.

Giving a general overview of the geology of Kivu and Maniema, tectonics controls the structure in a particular way in these two regions.

Recent periods have been marked by extension, reflected in the subsidence along the trench, the bottom of which is occupied by large lakes.

The subsidence of the bottom of the graben was accompanied by the raising of its banks, exposing them to intense erosion that rejuvenated the relief, thus clearing the oldest Proterozoic terrain in Central Africa. The beginning of the fourth millennium, in the east of the Democratic Republic of Congo, was characterised by a new distensive phase that rejuvenated the pre-existing faults.

These continental-scale fractures have rejuvenated the hydrographic networks in the graben zone.

Kivu is rich in mineral resources, most of which are Proterozoic: Fe, Sn, W, Nb, beryllium, REE, zinc, gold, silver, platinum, diamonds, emeralds, tourmalines, amethysts, geothermal energy, limestone and travertine, kaolin, methane gas and oil (Cenozoic and Quaternary). (Afazili Simba P., 2012).

Considering the geological assets of Kivu in general and probably Idjwi-Sud in particular, the whole of which has been hit by fairly major tectonic eras that affected the Kibaran.

Given Idjwi-nord's diverse lithology, it generally comprises metamorphic Precambrian rocks, plutonic rocks (pegmatites, granites, diorites) and quartz veins, schists, etc. These formations are altered by weathering, leading by diagenesis to Neogene rocks and sediments.

Recent tectonics throughout the continent have not spared the geological formations in the area, as can be seen from the partially altered diaclastic frictions.

The northern part of the Idjwi territory is dominated by granites that completely penetrate the Precambrian terrain, the oldest belonging to the Ruzizi system (KAVYAVU W., 2016).

Given the richness of Kivu and the geology of the northern part of the island, there is a serious need for in-depth research into the petrography and structure of the southern part of this territory.

Above all, this work allows us to consolidate the geological apprehensions of the NTAMBUKA chiefdom in the southern part of the Idjwi Territory, based on a reasoned study of its petrography and structure. It also constitutes a special prospecting guide if this region should later arouse curiosity in view of the existing clues which, once corroborated as a deposit, could contribute to the good of the Congolese nation.

As part of our study, we set out to answer the following questions:

- ✓ What types of rock are found at Ntambuka?
- ✓ What is its lithostratigraphic succession?
- ✓ What deformations (and structures) would be found in this environment?

To better conduct our study, we set ourselves the following objectives:

- ✓ A detailed survey of the geological formations of Ntambuka;
- ✓ Record structural information ;
- ✓ Draw up a geological sketch of the area.

We have therefore formulated the following provisional answers to the questions we asked ourselves:

- ❖ The southern part of the island is made up of the same types of rock found in the northern part (Bweshu and Kishumbu).
- ❖ The lithostratigraphic succession passes from micaschists to granite via lower or upper micaschists in which pegmatitic veins and

quartzites have developed; these characterise the Kibaran formations.

- It is also possible to find real deformations here, as tectonics greatly affected the Kibaran, causing breaks, folds, faults (or microfaults)...

II. METHODOLOGY

In order to achieve and satisfy the above objectives, we simply used descriptive, analytical and interpretative methods.

To help us carry out our fieldwork more effectively, we needed a range of equipment: field notebook and pencils, compass, clinometer, digital camera, GPS, hammer, sample bag, rucksack, magnifying glass, medical kit, etc.

- Collecting and selecting data in the field

The geological surveys consisted of locating outcrops, searching for them and collecting information from them. To do this, the routes were covered on foot and samples were selected on the basis of rock type and the different petrographic families. At each outcrop we made macroscopic observations (the structure and thickness of the layers, the degree of alteration, the colour, veins, fractures, joints, diaclases, minerals, etc.); the macroscopic observations of the samples enabled us to understand the structure of the rock, the colour, the types of minerals, the provisional name of the rock).

- Laboratory analysis of samples and data processing

Microscopic observation of the samples (petrographic analysis of thin sections) enabled the exact naming of the samples.

Geographical and structural data were processed on computer (using Excel and Word 2010 and software such as QGIS, Wintensor, Dips and Surfer).

III. DISCUSSION

3.1. PETROGRAPHIC ASPECT

Our study area is located in Idjwi territory in the Ntambuka chiefdom in the southern part of the island.

From a petrographic point of view, we can observe several formations, most of which are covered due to the fact that the soil has undergone considerable evolution. This explains why outcrops are rarer and vegetation occupies most of the region.

However, some outcrops are visible not only as a result of the various erosions observed, but also thanks to the flow of rivers along this sector. The area is said to be teeming with ancient volcanic formations linked to the evolution of the East African Rift, given the presence of lava in Kivu (presence of volcanoes). Plutonites are also to be found, hence the presence of granitoids in this region, and have undergone metamorphism, which is why gneiss, marble and schist can be found here...

As demonstrated by previous work (Benoît Smets, Damien et Al, 2016) and the work of (Gloire Bamulezi et Al, 2017), they showed in their survey that the Idjwi territory is made up of formations of the Ruzizian chain (Paleoproterozoic) from north to south, intersected by the Kibarian chain (Mesoproterozoic) in the centre of the island and volcanic rocks on the south-western part of the island.

With regard to our contribution to the petrographic study of Ntambuka, given that we have hypothetically put forward the argument that the southern part of the island is made up of the same types of rock as those found in the north, we need to confirm that these are the same types of rock from the genetic point of view (sedimentary, metamorphic and magmatic).

But in detail, by referring to the macroscopic and microscopic data, we find sedimentary rocks in small part in the region especially at the edge of Lake Kivu, sands resulting from the alteration of granitoids, kaolins, silicified limestones confused macroscopically with marble; volcanic rocks in abundance, pyroclastites; metamorphic rocks such as : schists, micaschists; plutonites such as: granites, pegmatites, quartzite diorites and intermediate rocks such as: microgranular diorites, quartzite microdiorites, dolerites.

3.2. CARTOGRAPHIC ASPECT

A geological map is a representation of the rocks and geological structures present in the outcrop or subsurface of a region.

Its aim is to show the spatial distribution of lithological facies, their succession, and the various tectonic structures. These maps usually ignore recent surface formations, concentrating on the underlying bedrock.

Following macroscopic descriptions and plotting of the various stations on a topographic base, we were able to obtain a lithological map.

In comparison with the results of previous mapping work carried out in Bweshu and Kishumbu on the northern part of the island by (KAVYAVU W., 2016), who used a scale of 1:20,000, this enabled him to map lithological units such as Granites and Pegmatites, which predominantly outcrop, shales, conglomerates, sandstones, sands and Orthogneiss.

Based on its mapping, we can see that the northern part is dominated by acid plutonites and detrital rocks can be seen at the edge of the lake.

But with our 1:100,000 scale used in our mapping, we were able to understand that volcanic rocks are predominant in the south-western part of the region, with the presence of pyroclastites and on this part of the island there are domes of ancient volcanic craters and calderas; this suggests that the island of Idjwi is partly a volcanic island.

Plutonic rocks, especially acidic ones, are found to the north of our study area, which explains the continuity of these formations found by our predecessors in the northern part of the area.

If you look at the lithological map of the region, you can see an almost repetitive succession of granite-schist-pegmatite; but given the scale used, which is too large, this is hardly noticed, whereas it would be very visible at a scale of at least 1/5000.

Towards the north-west of our study area, on the edge of the lake, we find conglomerates that do not outcrop considerably, so they are not shown on the map given the scale used.

The metamorphic rocks found in the region are mainly gneiss, schist and marble, which shows that granitisation also plays an influential role in the formations found there. And this justifies the Ruzizian and Kibaran characteristics.

3.3. STRUCTURAL ASPECT

Compared to previous studies carried out by Gloire Bamulezi et Al in 2017 on "Seismotectonic analysis of some potentially active faults in the western part of the Kivu rift in the Democratic Republic of Congo (DRC)"; in the regional setting of the northern Lake Kivu basin and the western part of the Virunga volcanic province. By taking structural measurements of the orientations of tectonic markers (faults with associated striations, tension joints, various fractures, etc.), the stress tensor obtained shows a horizontal extension oriented WNW-ESE.

This orientation is similar to that given by STAMPS et al (2008), DELVAUX & BARTH (2010) and FERNANDEZ et al (2013) for the Kivu rift as a whole. A main boundary fault zone (NNE-SSW) whose tectonic markers (striated fault planes, tension joints, various fractures, fault breccias mixed with clays) have been described at Muranga and Bweremana to the north of Lake Kivu, has been identified and characterised.

A series of light to moderate earthquakes recorded in January 2002 are thought to be associated with the reactivation of this boundary fault. These volcano-tectonic earthquakes were generated by the relaxation of extensional tectonic stresses following the intense volcanic activity of the Nyiragongo volcano on January 17, 2002. As for the earthquakes recorded on 24 October 2002, their seismotectonic parameters are integrated into the northern part of the Lake Kivu basin.

The southern zone includes the southern part of the Lake Kivu basin, the upper Ruzizi basin and the extreme north of Lake Tanganyika. The stress tensor shows a horizontal extension oriented almost E-W.

These are the faults described at Funu and Karhale (Bukavu "microrift") and at Kavimvira and Kalundu (North Tanganyika rift). The faults of the Bukavu "microrift" overlap, creating a sort of relay ramp within which surface deformations and landslides affecting infrastructure are observed (deterioration of roads, drains and gutters, cracking of houses, etc.). (Gloire Bamulezi et al, 2017).

Still within the regional framework, with reference to the studies carried out by SHUNGU L. Guy in 2018 on the Structural study of the formations of Western Lake Tanganyika (case of Kavimvira, DR Congo)

The microstructural study has enabled it to highlight planar structures such as schistosities and foliations, and discontinuous structures such as joints, diaclases, veins and detachments. Continuous structures were mainly observed on a regional scale.

He concluded that the schistosity planes have a NNW-SSE preferential direction and the foliations have a NNW-SSE preferential direction; the diaclases have an ENE-WSW preferential direction while the joints have a NNE-SSW preferential direction; and the veins have a NW-SE preferential direction.

Structural analysis of the conjugate fractures encountered in his study area shows that these fractures are derived slightly from extension and largely from compression resulting from stresses σ1, σ2 and σ3 whose values are σ1: N14°E/6°NNE, σ2: N99°E/41WNW and σ3: N112°E/49°ESE.

Based on the work carried out by KAVYAVU W. in 2016 on the characterisation of the deformation of the bweshu and kishumbu geological formations of Idjwi (Kibaran) in South Kivu, DR Congo, he concluded that

The structural approach to Idjwi in its northern part indicates that the schistosity plane is rare and is oriented around N45°E with a SE dip, flexible deformation is absent.

After processing the data, he found three structural networks at Bweshu, the first network runs from 30° to 80°, the second from 80° to 160° and the last from 160° to 180°. The preferred direction is N109° E/83°.

At Kishumbu, on the other hand, processing of the structural data reveals two major structural networks, the first ranging from 70° to 120°, the second from 140° to 200°, with a preferred direction of N147°/79° SW.

For our contribution to the structural study of Ntambuka in the southern part of the island, we understand that the continuous deformations are almost impossible to find, with the schistose planes orientated around NE-SW.

Processing of the structural data on the diaclases gives a single structural network running from 140° to 160°, while the preferred direction is N335°E/48°ENE. Structural analysis of the conjugate fractures encountered in our study area shows that these fractures result slightly from extension and largely from compression resulting from stresses σ1, σ2 and σ3 whose values are, σ1: N106°E/0°WNW, σ2: N15°E/79°SSW and σ3: N196°E/11°NNE.

As for the filled breaks, data processing led us to obtain a preferential direction of N190°E/55°WNW. Then the stress tensor gives us a distension zone with σ1: N166°E/68°NNW, σ2: N352°E/22°SSE and σ3: N261°E/2°ENE.

CONCLUSION AND OUTLOOK

Our research has shown that the geological formations in the Ntambuka chiefdom are typical of the Congolese Paleoproterozoic known as the Ruzizi sub-group, and are more or less comparable to those in the northern part of the island, although they differ in that volcanic rocks occupy an estimated 40% of the south-western region.

The lithological units, granites and pegmatites that cut through the schists, are seen towards the northern part of the chiefdom, but given the scale used in our mapping, this is hardly noticed. Towards the north-west, at the edge of the lake, we find conglomerates that outcrop very little. Limestone is also present, with some metamorphism, and intermediate rocks such as dolerites are also abundant, along with dioritic rocks.

After processing the data with WIN TENSOR software, it was observed that continuous deformation is rare, with the planes of the schistosities orientated around NE-SW. Open fractures have two preferential orientations NW-SE and NNW-SSE. The veins have a preferential NNE-SSW orientation.

The structural analysis of the conjugate fractures showed that the breaks are the result of compression resulting from stresses σ1, σ2 and σ3 whose values are, σ1: N106°E/0°WNW, σ2: N15°E/79°SSW and σ3: N196°E/11°NNE. Then that of the veins showed distension with σ1: N166°E/68°NNW, σ2: N352°E/22°SSE and σ3: N261°E/2°ENE.

The seismotectonic and microstructural data will be the subject of future work to understand the context of compression at the centre of a rift, which would appear to be an anomaly.

REFERENCES

1. AFAZILI SIMBA. L. PASCAL, 2012. Genèse, Structure et Minéralogie des Gisements Stanno-Wolframifères de la Ceinture Mobile de l'Est de la République Démocratique du Congo ; Faculté de Sciences ; Université de Lubumbashi.

2. BENOIT SMETS, DAMIEN DELVAUX et Al, 2016. The role of inherited crustal structures and magmatism in the development of rift segments: insights from the Kivu basin, western branch of the East African Rift. Tectonophysics 683 (2016).62.76; 15p

3. FOUCAULT. A, RAOULT.J-E, 2000. Dictionnaire de Géologie Paris, Masson édit, 5th Edition.

4. GLOIRE BAMULEZI, DAMIEN DELVAUX, et Al, 2017. Seismotectonic analysis of some potentially active faults in the western part of the Kivu rift in the Democratic Republic of Congo (DRC). In Geo-Eco-Trop, 2017, 41, 2, n.s.169-186

5. ILUNGA LUTUMBA, 1991. Morphology, Volcanism and Sedimentation in the South Kivu Rift. In Bull. Serv. Geol. Liège, 27: 209-228.

6. KAVUKE, MAHINDA et Al, 2012. À propos des Mouvements Tectoniques de l'île d'Idjwi dans le lac Kivu tel que révélé par les anomalies magmatiques de 1900 à 2010. 17p. In Observatoire Volcanologique de Goma, Département de Déformation GECOTEC/Bukavu Cahiers du CERUKI, Nouvelle Série, 42, pp. 26-42

7. KAVYAVU KAMBALE WISDOM., 2016. Characterisation of the deformation of the Bweshu and Kishumbu geological formations of Idjwi (Kibarian) South Kivu, DR Congo. 14p. In Conservation et Société, Vol. 1, N° 009 Sciences et sciences appliquées Mars 2016

8. POUCLET, A., 1975. History of the Great Lake of Central Africa. Mise au point des connaissances actuelles. In Rév-Géogr.Phys-Géol.dyn, 2, 17, 5: 475-482.

9. SHUNGU GUY L., 2018. Structural study of western Lake Tanganyika formations (case of Kavimvira, DR Congo). In International Journal of Innovation and Applied Studies; ISSN 2028-9324 Vol. 24 No. 1 Aug. 2018, pp. 299-310 - 2018 Innovative Space of Scientific Research Journals

10. www.ambardcprague.com

11. www.m.wikipedia.org

Table of contents

GEOCHIMICAL AND GEOPHYSICAL METHODS FOR THE PROSPECTION OF Au-Ag-Pt and U

CONTRIBUTION TO THE PETROGRAPHIC AND STRUCTURAL STUDY OF NTAMBUKA/IDJWI (SOUTH KIVU, D.R.C.)

MIX
Papier aus verantwortungsvollen Quellen
Paper from responsible sources
FSC® C105338

Printed by Books on Demand GmbH, Norderstedt / Germany